Teach Your Kids About

Nanotechnology

By Mark E. Tomassoni and
Rama Ramesh
Illustrated by Yoko Matsuoka

MARK E. TOMASSONI

is the creator and author of the *Nanobots for Kids* series designed to educate and entertain readers about nanotechnology. He lives and works in Laurel, Maryland, USA.

YUKO MATSUOKA

has provided professional graphics and illustrations since 2009 and has illustrated such works as *The Garden*, *Fiona Thorn*, and *Maniu and the Prince of Bakara*. She lives and works in Akita, Japan.

RAMA RAMESH

is a creative writer specialized in writing for children. She has written stories and non- fiction for many children magazines and books. She lives and works in New Delhi, India.

Image Credits

What Is Nanotechnology?

Nano and Nana are students just beginning to learn about nanotechnology—like you! Join them to find out what it's all about!

Everyone is talking about nanotechnology! What does "nano" mean? What is so special about nanotechnology?

The word "nano" means "one billionth." So a nanometer is only a billionth of a meter. "Can you imagine that?" Nano asks.

"Only atoms and molecules are that small and they make up literally everything in our universe—the stars, planets, trees, animals, birds, rocks, water, and even you!" says Nana.

"Nanotechnology is a science that deals with studying atoms and molecules and building nanomaterials to help people throughout the world," Nano says.

"All types of students are learning more and more about this technology. So it is important that everyone has a chance to see what the future can become," Nana adds. Since atoms and molecules are so tiny, it is impossible to see them with our eyes or even through an ordinary microscope. So scientists invented advanced instruments and powerful electron microscopes like the **scanning tunneling microscope (STM)** and the **atomic force microscope (AFM).**

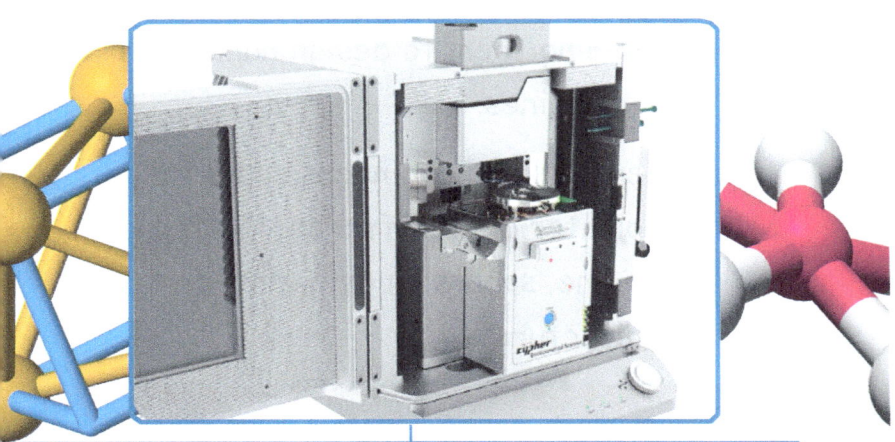

Powerful electron microscopes that are used to see molecules and atoms in nanotechnology.

This is how nanotechnology works:

- Scientists study atoms and molecules and learn how they behave.

- These scientists then look into the future to see what technology will become available and what can be created with the new *nanorobots* or *nanobots*.

- Scientists manipulate molecules and atoms by moving them around and combining them to design nanomaterial and machines.

- Scientists and engineers apply the nanomaterials in different fields like medicine, sports, engineering, and architecture.

In the future, nanobots will travel through our blood like this!

4

"And just think," Nano exclaims, "someday soon molecular-sized nanobots will travel through our bodies and be able to detect diseases or problems."

"We will be able to make body armor that is super strong and able to protect us from bullets or even cannons—we can become like Iron Man!"

Nana adds, "Televisions and computers will be rolled up and carried around in our pockets! Some nanobot computers will be located inside your eyes so you can see information without actually looking at a computer."

"I can't wait until nanobots are found in my sporting equipment like soccer balls or basketballs," says Nano. "Imagine nanoballs that can bounce ten times higher than a normal ball. We'll have to invent entirely new games!" All this and many more useful and novel inventions will be possible because of nanotechnology!

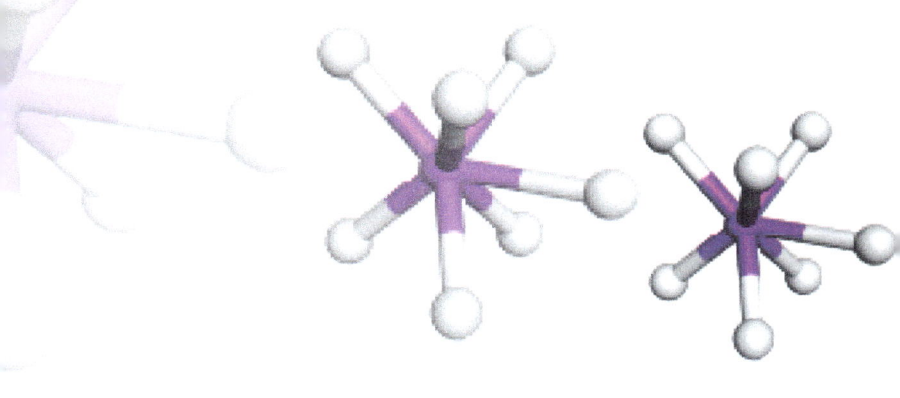

Getting Down to the Nanoscale

So really, what is nanotechnology? Well, **nanoparticles** are impossible to see and are 1,000,000 times smaller than the period at the end of this sentence. To understand how small nanoparticles are, think about a typical baseball bat being about 36 inches (one yard) or 100 centimeters (one meter) in length. One nanometer is one billion times smaller than the baseball bat! "Whew... that sets the brain reeling," says Nana.

Think of something else that is smaller. An ant? How about an ant's eye? "A nanoparticle is still a million times smaller than any of these!" says Nano.

"Look at one strand of your hair," Nana says. "Look at the tip—you can hardly see it. It is tinier than a pinhead, right? Do you know that the diameter of one strand of hair itself is 50,000 nanometers wide?"

QUESTION: How thick is a sheet of paper?
ANSWER: About 100,000 nanometers! That's a lot of nanometers.

Fun Fact

Did you know that your fingernail grows one nanometer each second?

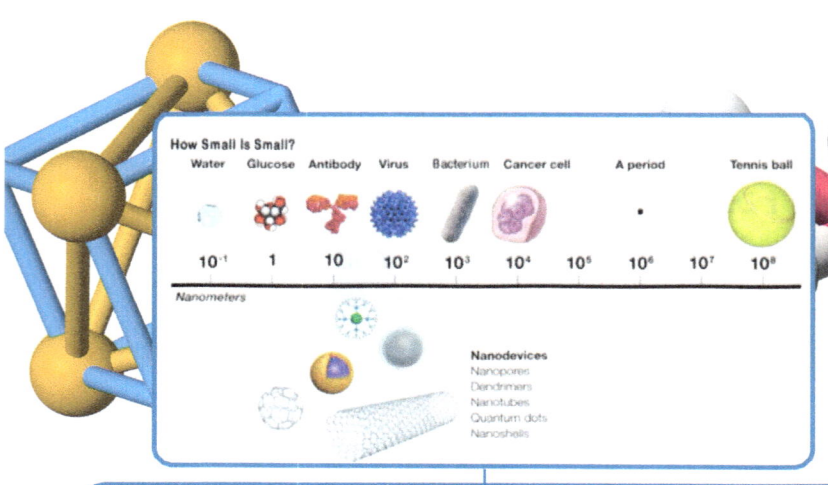

How Small Is Small?

Water	Glucose	Antibody	Virus	Bacterium	Cancer cell		A period		Tennis ball
10^{-1}	1	10	10^2	10^3	10^4	10^5	10^6	10^7	10^8

Nanometers

Nanodevices
Nanopores
Dendrimers
Nanotubes
Quantum dots
Nanoshells

Nanoscale. Compare the sizes.

Nanotechnology: A Brief History

"This book is teaching me a lot about nanotechnology," Nano and Nana say. "Here is some of the early history that is important to understand in order to plan for the future."

Early Research. In 1959, the American physicist, Richard Feynman, proposed ways to work with individual atoms and molecules. Most people thought his ideas were crazy and nobody thought that such things would be possible!

But a few years later an American engineer, Eric Drexler, came up with a scary concept called "grey goo." He proposed that in the future, tiny nanobots would get out of control and destroy everything on earth.

Now, thank goodness, most scientists do NOT agree with Drexler's idea. But, certainly, working with nanomaterials requires highly trained personnel using great care.

Electron Microscopes. But it was the invention of powerful microscopes that really allowed Feynman and Drexler to see the potential of nanotechnology. Now the world needs more and more young engineers and scientists who can work with nanobots to better people's lives around the world. Will you help us? "We really need you," Nana and Nano say.

Timeline of Major Nanotechnology Events

1959. Richard Feynman is the first person to describe the concept of building molecular machines.

1974. The term "nanotechnology" is used for the first time.

1981. *The scanning tunneling microscope is invented, helping scientists view molecules in nanoscale.*

1985. *The buckminsterfullerene, also know as the buckyball structure, is developed.*

1989. *IBM spells out its logo using individual atoms.*

2011. *First programmable nanowire circuits are used for nanoprocessors.*

And in the near future...nanoparticles will deliver medicines and cancer treatments in the body. Better cures and longer lifespan is possible with nanotechnology!

Fun Fact

The scientists who developed the buckyball, which was believed to offer tremendous potential for future projects, received the Nobel Prize for Chemistry in 1996.

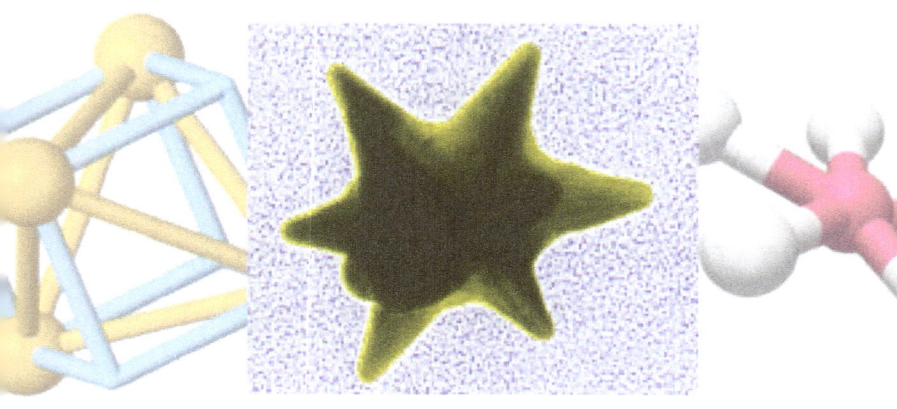

Nanomaterials

"Nanomaterials," says Nana, "come in different shapes like rods, spheres, and films made of molecules or tiny nanoparticles." Once a nanostructure is constructed or assembled, scientists try to understand how they behave. For example:

- Is the nanostructure magnetic?
- Is it transparent or opaque?
- Is it a good or bad conductor of heat?
- How does it react with different chemicals?
- Does it conduct electricity?
- How strong or weak is its structure?

After answering these questions, scientists gain a good idea about how to use a particular nanomaterial for a particular purpose. For example, a very strong nanomaterial can be used for police or military body armor. And new materials are being considered for automobiles to make them lighter and also stronger. New types of concrete are also being studied for building homes or roads.

Can you think of other ways nanomaterial can be used?

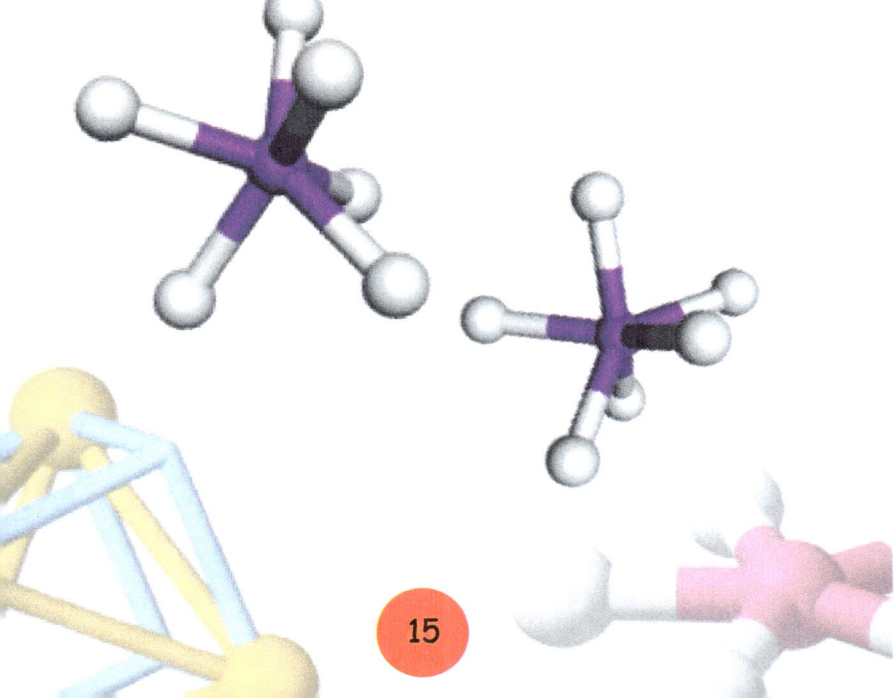

Fun Fact

Scientists have developed carbon nanotubes that are extremely strong and conduct heat faster than any other material we know!

Nano Products We Can Buy!

Who said nanotechnology products are a thing of the future? Nano and Nana say that these nanoproducts are available already. Have you bought any? Do you know people who use any of these products?

1. Moisturizers, face creams, and sunscreen lotions contain nanoparticles.

2. Bats, balls, and other sports equipment made of nanocomposites.

3. Yacht sails made of nanofabric that is waterproof and gathers more wind.

4. Anti-bacterial socks coated with silver nanoparticles that help contain bad odor!

5. Nano-enhanced clothing that are stain resistant and waterproof.

6. Nanomaterial composed of carbon nanotubes that are strong, yet lightweight, and used in building masts for ships.

7.   Paints that contain nanomaterials that are mold resistant, easy to clean, and that last longer!

Nanotechnology in Progress

Nano and Nana are really lucky since they get to use special nanotechnology products before other kids can. Find out from Nano and Nana what products they can use now and what products will be available in the future!

Nano Clothing

Nano is going out, and he'll be wearing a special T-shirt coated with titanium nanoparticles.

What do you think it's useful for?

This T-shirt will prevent harmful ultra- violet (UV) rays from the sun entering through the T-shirt and harming his skin.

Nana has clothes coated with nanomaterial that absorbs sweat and remains dry for a long period. She has a shirt coated with silver nanoparticles that kills bacteria that causes body odor.

Nano no longer worries if he spills milkshake on his shorts. Do you know why? His shorts have a nanofiber coating that stops liquids from seeping through the fabric. Really cool.

Clothes with useful features can be designed using nanotechnology!

Nano Solar Power

Nana says her dad is using nanomaterials in solar cells to help improve solar energy production.

How is this possible?

Nanomaterials coated on the solar cells can harness sunlight more effectively. More sunlight means more energy! Also, some parts of solar cells are expensive, but they can be replaced with cheap nanomaterials. Nano is using nanomaterial films to help him convert solar energy to electricity.

Better solar power with the help of nanotechnology.

Can nanotechnology even make batteries last longer? Nana says yes! She got some batteries for her video player remote control.

And it's been two years, and the batteries are still fine. The secret? The battery electrodes are coated with a type of nanoparticle that protects the batteries like shields. This way, the battery doesn't get damaged and will last for a long time.

All Nana has to do is recharge them again and again!

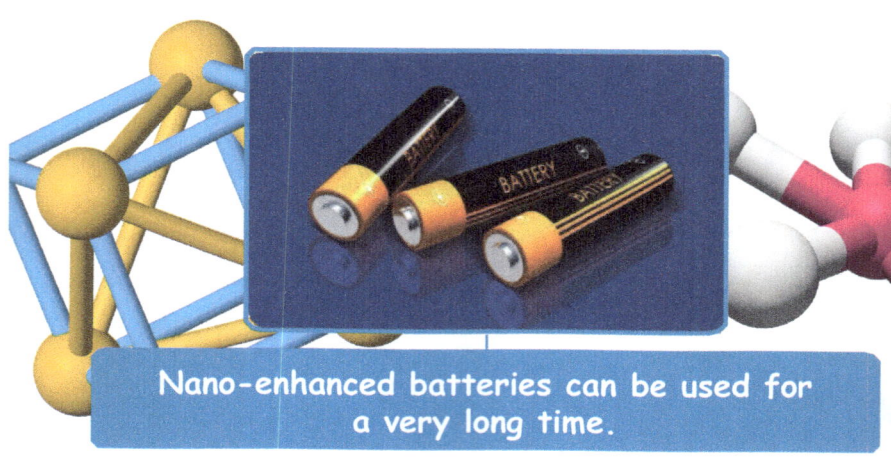

Nano-enhanced batteries can be used for a very long time.

Nano Computer Chips

Nano downloads movies in a few seconds; this is possible because his computer's processor is using nanoscale computer chips, which makes the computer run faster and helps it store more information.

Nanoscale computer chips for faster and more powerful computers.

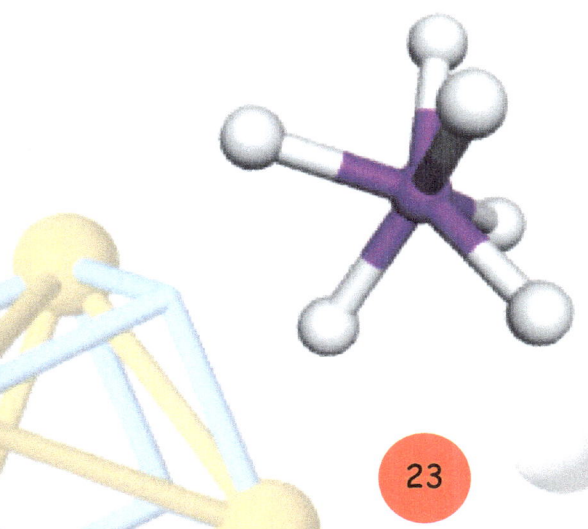

Nano Sporting Goods!

Think of a baseball bat that swings better or a tennis racquet that can hit the ball more forcefully! "Special nanofiber material give bats or balls extra strength," Nano says. He's going to use the baseball bat his dad gave to him. Stronger and better sports goods, thanks to nanotechnology.

It's not just better bats and balls. Nano uses gloves and pads that are light as a feather,

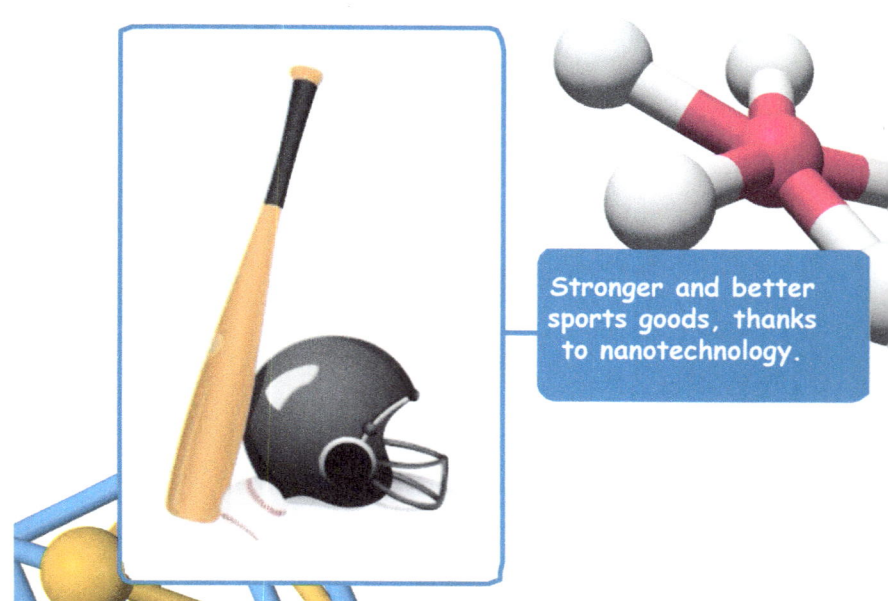

Stronger and better sports goods, thanks to nanotechnology.

thanks to special nanomaterial. Wouldn't that make it easier for professional sports players?

Vrooom... Nanotechnology

Nano and Nana's family car has no windshield wipers, but it has something better than that to keep raindrops away from the windshield! What? you ask. Nano says the windshield is coated with thin nano-films that repel water. Nano says that his dad is planning to design tough nanomaterial to coat his car so that it won't get scratched. This will be useful for all car owners.

Cars with better and exciting features using nanomaterials.

Car bumpers, cargo liners, and other vehicle components can also be designed with tough carbon nanofibers and nanotubes. This will prevent dents in vehicles even upon collision.

Making Food Better with Nanotechnology

Ever seen food labels with the words, "Best before such-and-such a date," or "Consume within 3 days," or something similar?

With the help of nanotechnology, it is possible to increase the shelf life of packaged food stuff.

Nana has packets coated with nanocomposite material that makes food last much longer than usual. She says that kitchen utensils like spoons, forks, and spatulas and the kitchen countertop in her house are very clean because they are coated with silver nanoparticles that kill bacteria and fungi.

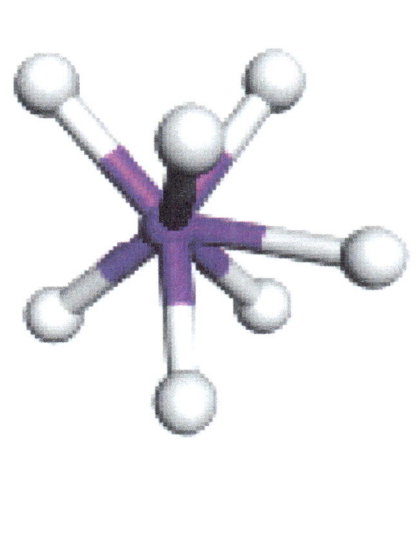

Food can be stored for a long period using nanomaterial packaging.

Exciting Future Prospects

Nanotechnology for Better Homes

Nano, Nana, and their parents are thinking of exciting possibilities in the future such as building ultra-modern houses with the help of nanotechnology.

Think of a house with thin but protective walls made of tough nanomaterials and specially treated nanoparticle-coated windows that can regulate the amount of light and heat that enters the house.

Nanomaterial for ultra-efficient homes.

Such a building will have solar panels embedded throughout its exterior surface. This house will have enough energy to run all appliances within it. Imagine concrete and laminates that are capable of storing energy through all weathers and seasons.

Consider a house equipped with nanofilters that recirculates its own water. Consider walls, windows, and roofs treated with durable nanoparticle coatings that can break down pollutants, clean the air, and remain free from mould, dirt, and dust.

Imagine a building that can adjust to your mood. Your emotions will control the lighting, ventilation, and sound through sensitive detectors.

Imagination is the only limit to the potentials of nanotechnology!

Nanobot and Molecular Manufacturing

Have you seen an old movie called *Fantastic Voyage?* In this movie, a submarine and the crew inside are shrunk and injected into the body of a sick man so that the crew could examine what's wrong and save him. Wouldn't it be wonderful if such a thing were really possible? Nano and Nana say this is definitely possible in the near future!

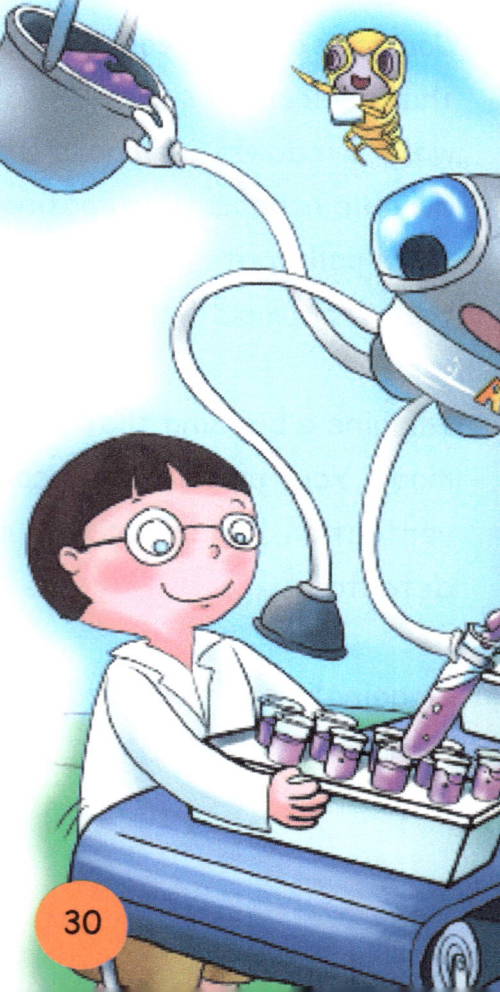

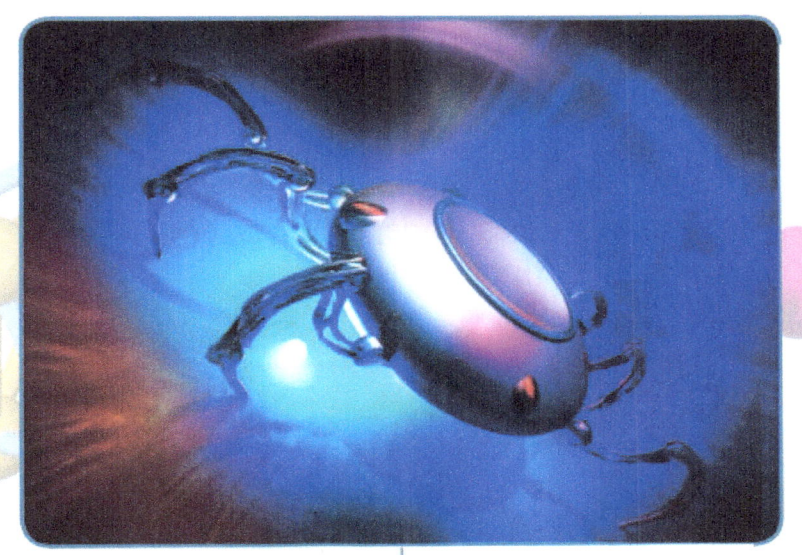

In the future, nanobots like this can help doctors perform surgeries with greater accuracy.

Scientists are developing tiny nanobots to explore the inside of a body and perhaps kill cancerous cells or harmful bacteria!

As their name suggests, nanobots are equipped with moveable arms and other features depending on their role.

The possibilities and uses of nanobots are endless. They can help soldiers spy on enemies. They can help repair tiny computer chips. They can monitor pollution and inform us of dangerous chemicals in the air or water.

Molecular Manufacturing

Molecular manufacturing is the process of constructing wires, rods, spheres, and other structures at the nano level.

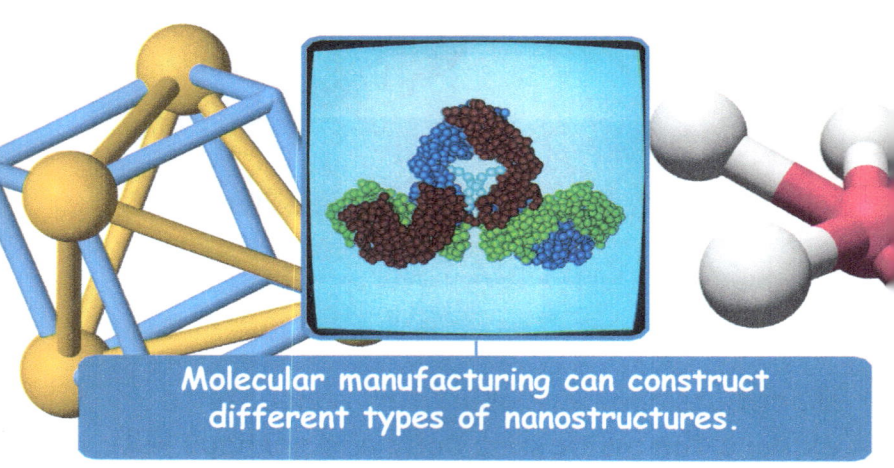

Molecular manufacturing can construct different types of nanostructures.

Better Security with Nano

Nanocomposite material can be used for designing lightweight body armor with immense strength and ability to survive even heavy impact! Nano is even thinking of a "liquid armor" that will be made of flexible nanomaterials capable of molding to a person's body.

Body armor developed with nanomaterials can even protect persons from cannons!

Saving Lives with Nano

Nana says that one of the biggest challenges for physicians is the need to know and treat diseases in their early stages. Many side effects of taking drugs can be reduced if the drugs are delivered only to specific cells. Nano vehicles will be capable of delivering the drugs to specific locations within the body.

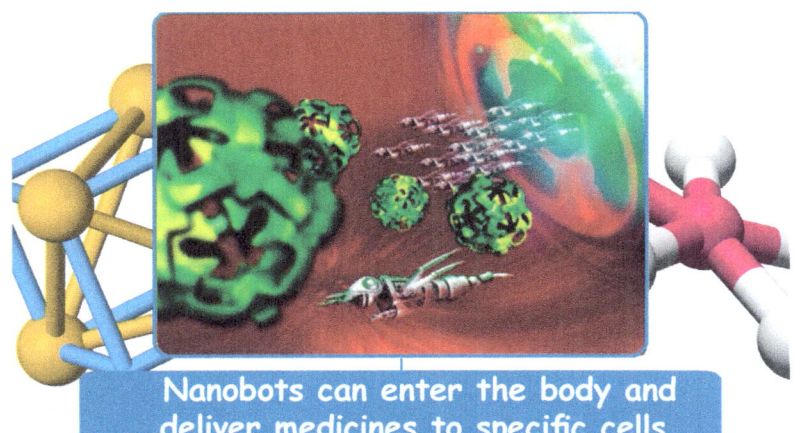

Nanobots can enter the body and deliver medicines to specific cells.

Think of the buckyball. Wouldn't it be a good idea to pack medicine within its cage-like structure and inject it into a sick person?

Specialized nanoparticles that stick only to cancer cells can be designed. Those cells alone can be targeted and destroyed!

Nanotechnology, Generation Next

Thanks to nanotechnology, it will be possible to have super fast processors; computers that are lightweight and smaller have enormous storage capacity and consume less electricity.

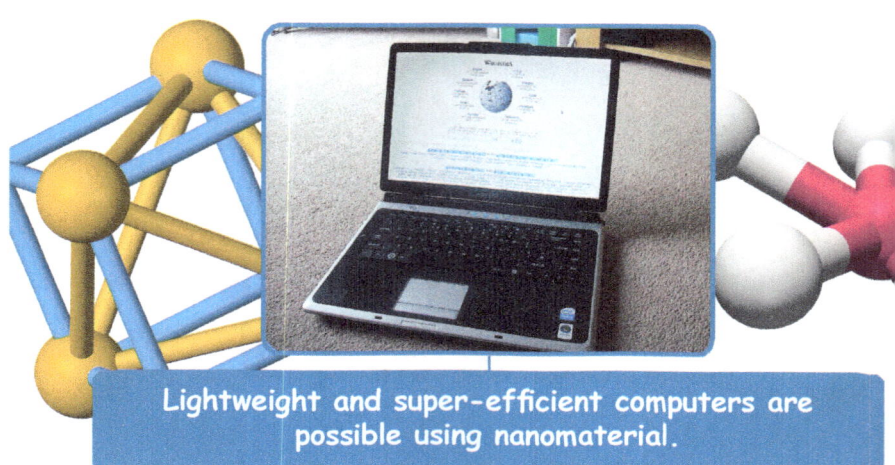

Lightweight and super-efficient computers are possible using nanomaterial.

Nana says that in order to accomplish this, computer processors and their electronic parts have to be made of specific nanomaterial.

A Cleaner and Greener Planet

Nanotechnology can help our Planet Earth in many ways, such as making solar cells more efficient in capturing the sun's rays and converting them into energy. As you know, solar energy is clean and causes no pollution. Batteries and fuel cells can be recharged to be used again for long periods and therefore do not need to be thrown away so quickly.

Automobile exhaust pipes can be coated with nanomaterials to absorb carbon dioxide and toxic chemicals. This will reduce water and soil pollution.

It's up to us to protect our planet and make it greener!

By injecting nanoparticles into the soil, they can bind contaminants, making for cleaner and safer groundwater.

Nanomaterials can even replace harmful materials that pollute the environment.

Nano says, "We kids should get together and think of many other things we can do with the help of nanotechnology.

So are you ready?"

Nano Quiz

How well do you know nanotechnology?
Test yourself!

1.   A nanometer is just one ___ of a meter!

    a.   Hundredth

    b.   Thousandth

    c.   Millionth

    d.   Billionth

2.   Imagine a nanoparticle that is 100 nanometers wide. Will you be able to see it with your eyes?

    a.   Yes

    b.   No

3. Physicians are already using nanobots to kill bacteria inside human bodies.

   a. True
   b. False

4. Silver nanoparticles are used for killing bacteria that cause bad odor.

   a. True
   b. False

5. Nanoscale thin films that repel water are best suitable for which of the following:

   a. Windshield wipers
   b. Water faucets
   c. Dishwashers
   d. Computer chips

6. A scientist wants to design a nanobot for spying purposes. Which of the following will be the most useful feature for this nanobot?

a. Embedded solar cell
b. Inbuilt camera
c. Bendable legs
d. Lightweight body

7. A company is producing a tiny camera about as small as a pin's head. Is this a nanoscale product?

a. True
b. False

8. An artist is going to paint on a huge 20-foot banner, which will require lots of paint. What would be the most useful nanotechnology product for him?

a. Spy nanobot
b. Anti-odor socks
c. Stain-resistant clothes
d. Nano-enhanced tennis racquet

Answers: 1-d, 2-b, 3-b, 4-a, 5-a, 6-b, 7-b, 8-c

Glossary

Assembler: Micro-machine that will provide billions of nanomachines each day for different purposes.

Buckyball: A three-dimensional structure that is made up of exactly 60 atoms and resembles a football.

A buckyball is spherical and looks like a cage.

Molecular manufacturing: The process by which atoms and molecules are assembled to form a nanostructure.

Nanobots: Nanoscale robotic devices with different features designed for various purposes like medicine delivery, surveillance, etc.

Nanocomposite: A complex nanostructure made of different nanomaterials.

Nanomaterial: Material filled with nanostructures. Think of a material filled with plenty of spaces between the molecules and similar to foam.

Nanometer: One billionth of a meter, written as nm.

Nanoparticles: Tiny three-dimensional (3-D) particles, measuring one to only a few nanometers.

Nanorods: A nanoscale structure that is shaped like a rod. Nanoscale Film: An ultra thin film measuring only a few nanometers and very, very thin!

Nanosphere: Nanoscale structure that is spherical in shape.

Nanotube: A long, two-dimensional structure made of many atoms. Nanorods and nanowires are also just like nanotubes.

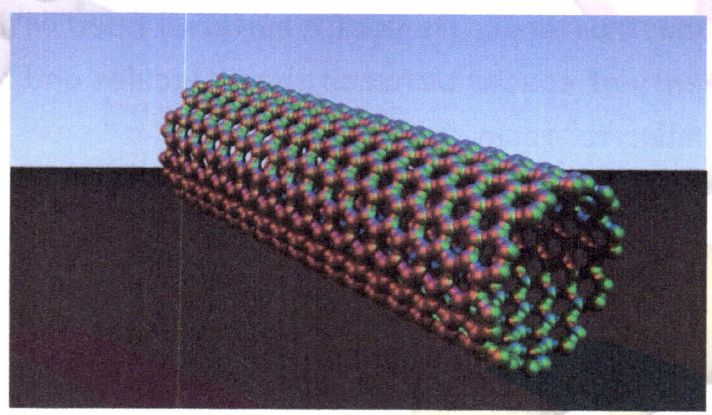

A nanotube made up of many atoms.

Nanovehicles: A nanoscale structure designed to travel through the human body, mainly used for medical purposes like drug delivery and detection of any diseases.

Nanowires: Wire shaped structures in nanoscale, similar to nanotubes and nanorods.

Quantum dots: Dots two to ten nanometers in diameter that emit light at different color wavelengths.

Scanning tunneling microscope (STM) and **atom force microscope (SFM):** Powerful microscopes that scientists use for viewing and working with nanoparticles.

Additional Reading

Websites

NanoandMe.org

www.nanobotsforkids.com

Books

Allhoff, Fritz; Lin, Patrick; Moore, Daniel (2010). What is nanotechnology and why does it matter?: from science to ethics. John Wiley and Sons. pp. 3-5. ISBN 1-4051- 7545-1.

Berube, David (2006). Nano-Hype: The Truth Behind the Nanotechnology Buzz. Amherst, NY: Prometheus Books.

Drexler, K. Eric (1986). Engines of Creation: The Coming Era of Nanotechnology. Doubleday. ISBN 0-385- 19973-2.

Drexler, K. Eric (1992). Nanosystems: Molecular Machinery, Manufacturing, and Computation. New York: John Wiley & Sons. ISBN 0-471-57547-X.

Nalwa, H. S. Encyclopedia of Nanoscience and Nanotechnology, American Scientific Publishers. ISBN 1-58883-163-9.

Nanotechnology Today, January 23, 2009. Prasad, S. K. (2008). Modern Concepts in Nanotechnology. Discovery Publishing House. pp. 31–32. ISBN 81-8356-296-5.

Lightning Source UK Ltd.
Milton Keynes UK
UKOW06f0208270216

269213UK00010B/33/P